AF878437

INTRODUCTION

This Book Is Not About Picking a Job
You don't have to know what you want to be.
Not now.
Not next year.
Maybe not even when you're older.
And that's okay.
A lot of books try to rush kids into choosing a future.
This book doesn't.
This book is about something more important.
It's about how your mind works.
Some kids like taking things apart just to see what's inside.
Some kids ask "why?" so many times that adults run out of answers.
Some kids notice patterns. Some imagine new worlds.
Some like fixing problems. Others like protecting people.
Some love numbers. Some love nature. Some love building things that move.
None of those ways of thinking is better than the others.
They're just different.
And each one can grow into something amazing.
You don't need to "pick a career" right now.
That's not the goal.

The goal is to start noticing:

- What makes you curious
- What makes time fly
- What kinds of problems do you like solving
- What feels exciting instead of exhausting

That's how real paths begin.
This book isn't here to tell you,
"Here's what this job does."
Instead, it asks a better question:
"Does this sound like you?"

Each chapter introduces a different way curious people use science, technology, engineering, and math to help the world. But more than that, each chapter shows what it feels like to think that way.

You'll read about people who:

- Turn ideas into instructions
- Build things that move, fly, or think
- Protect others in ways you can't even see
- Ask questions no one has answered yet
- Fix problems humans created
- Invent things that don't exist yet

As you read, you might feel excited.
You might feel unsure.
You might feel like more than one chapter sounds like you.
That's normal.

Most people aren't just one thing.
You don't have to finish this book with an answer.
You don't need a plan.
You don't need to choose.

All you need to do is pay attention.
If a chapter makes you think,
"Hey... I do that,"
or
"That sounds fun,"
or
"I'd like to try that,"
That's not random.
That's your curiosity tapping you on the shoulder.
And curiosity is one of the strongest skills you can have.
So take your time.

Read with an open mind.
Imagine yourself in different places, solving different kinds of problems.
This isn't about your future job.
It's about discovering who you already are.
Let's explore!

TECHNOLOGY CAREERS (4)

1. Software Creator (Software Developer)

You Turn Ideas Into Instructions

The learning app freezes.
A few kids groan. Someone taps the screen harder.
The teacher sighs and says, "Let's try again later."
But you lean closer.

Not because you're annoyed—but because something feels interesting.

You start wondering:

- Where did it stop?
- Did it miss a step?
- What if one small thing is out of order?

You don't see a broken app.
You see a puzzle.
That moment—when your brain wakes up instead of giving up—is how software creators think.

This Might Be You If...

- You like figuring out rules and patterns
- You enjoy giving clear instructions
- You notice when steps are missing or out of order
- You don't quit right away when something fails
- You like testing ideas "just to see what happens"

You might be the kid who:

- explains games really well
- says, "Wait—let's try it this way instead"
- enjoys fixing things more than using them

That's not being picky.
That's software thinking.

What Software Creators Really Do

Software creators write instructions for computers.
Those instructions are called code.
Computers don't guess.
They don't "kind of understand."
They only do exactly what they're told.
That means:

- every game character that moves
- every button you tap
- every sound an app makes

was planned step by step by a human.

Software creators build:

- video games
- apps
- websites
- animations
- tools that help people learn, work, and communicate

When something doesn't work, they don't panic.
They debug.

That means they:

- look at the instructions
- find the part that went wrong
- fix it
- try again

Sometimes they fix ten tiny problems before one big thing works.
That doesn't bother them.
Because every mistake gives them information.

How It Feels to Think Like This

Thinking like a software creator feels clear and focused.
It feels like:

- turning a messy idea into simple steps
- watching something you imagined come alive on a screen
- feeling proud when a tiny fix solves a big problem
- getting curious instead of frustrated when things break

In this world, mistakes aren't embarrassing.
They're useful.
A broken program isn't a failure.
It's a clue.

Tools You Might Use (and Why They Matter)

Software creators use tools like:

- Scratch – great for learning through games and animations
- Python – simple to read, powerful to use
- JavaScript – makes websites interactive
- computers, tablets, and sometimes robots

These tools matter because software runs the modern world.

Software helps:

- doctors treat patients
- cars drive safely
- schools teach and connect students
- scientists explore space

Behind all of that is someone who enjoys thinking step by step.
Someone like you.

Try It Yourself – Solve With Steps .

Pretend a human is a robot.

Your job: write instructions for brushing teeth.

Not:

"Brush your teeth."

Every single step.

Example:

- Pick up a toothbrush
- Turn on the water
- Get the toothbrush wet

Now check:

- Did you skip anything?
- Could the "robot" mess this up?

Fix your instructions so the robot can't make a mistake.
That's exactly how programmers debug.

One Last Thing to Know

You don't need to:

- Be "good at computers" yet
- Know coding right now
- type fast

What matters most is how you think:

- curious
- patient
- willing to try again
- interested in how things work

Those skills grow.

If you like turning ideas into instructions, fixing puzzles, and watching your thinking create something real...

This world might feel like home.

You Might Also Like

Data Detective (Chapter 4) and AI Builder (Chapter 13)

2. Game Designer (Video Game Designer)

DID YOU KNOW

Every jump, reward, surprise, and "almost won!" moment in a game was planned on purpose.
Games don't feel fun by accident.
Someone decided:

- when the game should be easy
- when it should get harder
- when it should surprise you
- when it should give you hope instead of making you quit

That someone is a game designer.

If you've ever played a game and thought:

- This level is too easy
- That part was unfair
- I would change this

You weren't just playing.
You were designing in your head.

This Might Be You If…

- You care about how things feel, not just how they work
- You like stories, characters, and challenges
- You notice when games are fun—or frustrating
- You enjoy puzzles, rules, and making things fair
- You imagine new worlds in your head

You might be the kid who:

- invents games at recess
- changes the rules to make things more exciting
- explains how to beat a level step by step

That's not "just playing."
That's design thinking.

What Game Designers Really Do

Game designers don't just play games all day.
They create experiences.
They decide:

- what is the goal of the game is
- what the player can and can't do
- how hard or easy each part should be
- when the game should surprise you
- when it should challenge you

They design:

- characters
- levels
- rules
- rewards
- obstacles

A good game designer thinks about the player the whole time.

They ask:

• Is this fun or just confusing?
• Is this too easy or too hard?
• Will the player want to try again if they fail?

Game designers work with artists, programmers, writers, and testers.

But their main job is simple:
👉 Make sure the game feels right.

How It Feels to Think Like This

Thinking like a game designer feels playful—but focused.
It feels like:

- imagining yourself inside the game
- caring about other people's experiences
- adjusting things again and again until they're just right
- watching someone play and learning from their reactions

When something doesn't work, game designers don't take it personally.
They test it.
They change it.
They test again.
That's called playtesting.
And it's one of the most important parts of game design.

Tools You Might Use (and Why They Matter)

Game designers use tools like:

- paper and pencils for early ideas
- storyboards to plan scenes and levels
- game engines to build digital worlds
- playtesting tools to watch how players react

Games aren't just for fun.

They help people:

- solve problems
- work as a team
- make decisions
- stay calm when things are hard

Games can help people learn, relax, connect, and imagine.
That's a big responsibility—and a powerful one.

Try It Yourself – Design for Fun

Design a simple game on paper.

Decide:

- What is the goal?
- What are the rules?
- What makes it challenging?
- What makes it fun instead of frustrating?

Now imagine someone else playing it.

Ask yourself:

- Would they understand what to do?
- Would they want to keep going if they lost?

If the answer is "yes," you're thinking like a real game designer.

One Last Thing to Know

You don't need to:

- be an artist
- be a programmer
- have fancy tools

What matters is that you care about:

- fun
- fairness
- creativity
- how others experience what you make

If you love creating worlds, shaping challenges, and thinking about how people feel while they play…

This might be your kind of world.

You Might Also Like

Software Creator (Chapter 1) and Inventor & Prototyper (Chapter 16)

3. Digital Guardian (Cybersecurity Specialist)

You Protect the Invisible World

You have a secret.

It could be:

- a private message
- a photo
- a school account
- a game login

You want the right people to see it.

But here's the problem:
You can't lock it in a box.
You can't hide it under your bed. You can't stand guard next to it.
And someone you've never met might try to get in—quietly.
So what do you do?
If your brain starts thinking about rules, protection, and "what could go wrong," you're already thinking like a cyber defender.

This Might Be You If...

- You notice when something feels "off."
- You like solving mysteries and puzzles
- You enjoy protecting people or systems
- You think ahead and imagine what could go wrong
- You like finding weak spots and fixing them

You might be the kid who:

- makes strong passwords
- asks, "What if someone cheats?"
- enjoys games where you have to outsmart an opponent

That kind of thinking isn't suspicious.
It's protective.

What Cyber Defenders Really Do

Cyber defenders protect the digital world.

That includes:

- personal messages
- photos
- school records
- hospital systems
- bank accounts

Their job is simple to say—but hard to do:

👉 Let the right people in. Keep the wrong people out.
To do that, cyber defenders have to think like both:
- the protector
- and the attacker

They ask questions like:
- Where is this system weak?
- What could be guessed or tricked?
- How can we make this safer before something goes wrong?

Some cyber defenders even test systems by pretending to break into them.
This is called ethical hacking.
They don't cause harm.
They find problems early—so real attackers can't.
Their goal isn't to scare people.
It's to keep people safe.

How It Feels to Think Like This
Thinking like a cyber defender feels calm and alert.

It feels like:

- spotting patterns others miss
- feeling responsible for keeping things safe
- solving quiet problems before they become big ones
- winning without being seen

Cyber defenders don't need applause.
Most people never know what they stopped.
And that's the point.
When security works, nothing dramatic happens.

Tools You Might Use (and Why They Matter)

Cyber defenders use tools like:

- strong passwords and password managers
- firewalls that block unsafe access
- encryption that scrambles information
- security tests that look for weaknesses

These tools matter because so much of life is digital now.

- Hospitals need secure systems to care for patients
- Schools need to protect student information
- Families need their private data to stay private

Cyber defenders help keep trust alive in the digital world.

Try It Yourself – Solve for Safety

Think of a secret message you want to protect.

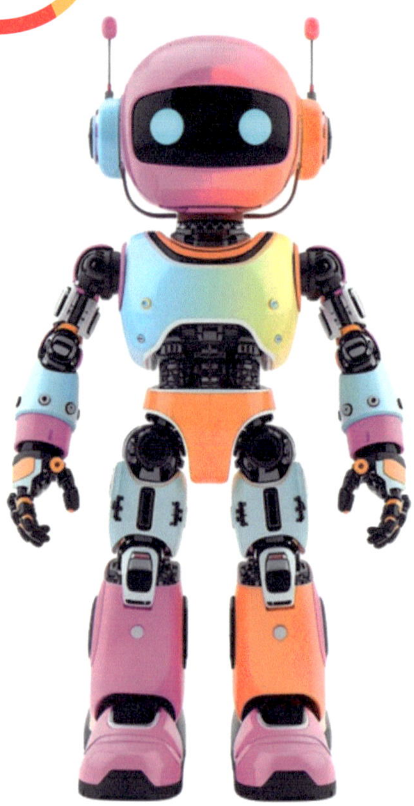

Ask yourself:

- Who should be allowed to read it?
- What if someone tries to guess it?
- How could you hide it better?

Try three methods:

- Change the letters
- Use a code only you understand
- Lock it behind a rule

Which one feels strongest?
That's cybersecurity thinking in action.

One Last Thing to Know

You don't need to be sneaky or suspicious to be a cyber defender.

Good cyber defenders are:

- careful
- ethical
- responsible
- curious

They use their skills to help—not harm.

If you like protecting others, thinking ahead, and solving quiet puzzles most people never notice...

This world might fit you perfectly.

You Might Also Like

Data Detective (Chapter 4) and AI Builder (Chapter 13)

4. Data Detective (Data Analyst)

You Find Stories Hidden in Numbers

There are pages full of numbers.
Charts. Lists. Scores. Times.
Most people see a mess.
You see a question.
What does this mean?
Why did this change?
What's hiding underneath all these numbers?
If you've ever looked at information and felt curious instead of bored, you might already be thinking like a data detective!

This Might Be You If...

You like sorting and organizing things
You enjoy comparing and spotting differences
You ask questions and want proof
You notice patterns other people miss
You like answers that make sense
You might be the kid who keeps track of scores.
Or the one who notices trends, like what day everyone is tired.
Or the one who asks, "How do we know that's true?"
That's not overthinking.
That's detective work.

What Data Detectives Really Do

Data detectives study information to understand what's really happening. Data can come from:

- Numbers
- Charts
- Surveys
- Measurements
- Records over time

Their job is to turn all that information into something useful.
They ask:

- What is happening?
- Why might it be happening?
- What could happen next?

Data detectives help people make smart decisions.
They might help:

- A school improves learning
- A city plans safer roads
- Doctors understand health trends
- Teams figure out what's working and what's not

Numbers don't speak on their own.
Data detectives help them tell their story.

How It Feels to Think Like This
Thinking like a data detective feels steady and thoughtful.

It feels like:

- Looking closely instead of guessing
- Feeling confident because you have evidence
- Enjoying quiet "aha!" moments
- Solving puzzles with facts instead of opinions

Data detectives don't rush.
They take their time because accuracy matters.
Being right is more important than being fast.

Tools You Might Use (and Why They Matter)

Data detectives use tools like:

- Spreadsheets to organize information
- Graphs and charts to see patterns
- Visual models to explain ideas
- Simple math to compare and measure

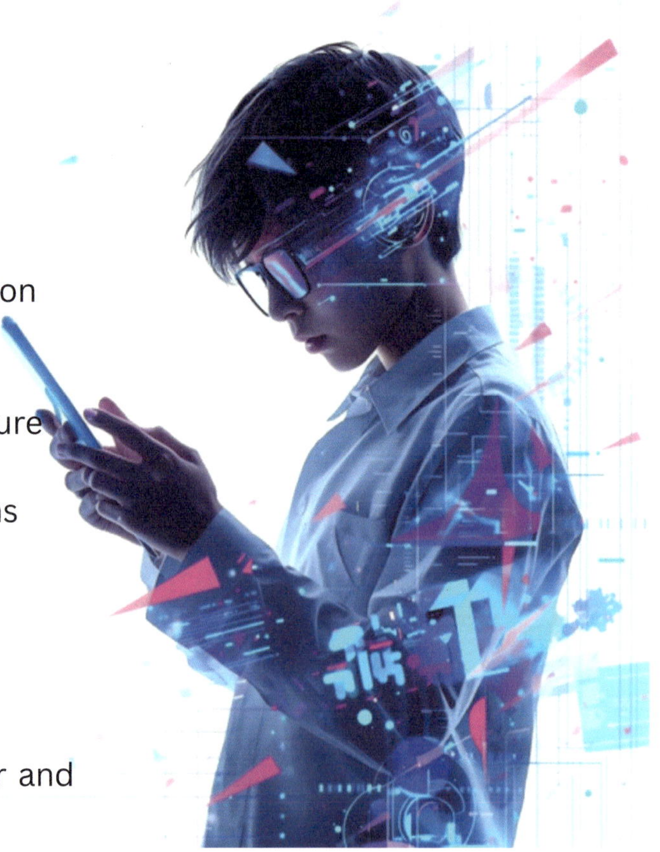

These tools matter because decisions affect real people.
Weather forecasts use data.
Health studies use data.
Sports teams use data.
Schools and cities use data, too.
Good data helps people plan better and avoid mistakes.

Try It Yourself

Pay attention to how you spend one afternoon.

Write down:

- How long have you been playing?
- How long do you read or study
- How long do you relax

Now turn that into a simple chart.

Ask yourself:

- What surprised you?
- What took the most time?
- What would you change tomorrow?

That's data detective thinking in action.

One Last Thing to Know

You don't have to love math class.

What matters more is that you:

- Like asking questions
- Want answers that make sense
- Care about being fair and accurate

If you enjoy finding patterns, telling the truth with facts, and helping others understand what's really going on...
This world might feel just right for you.

ENGINEERING CAREERS (4)

5. Robot Builder (Robotics Engineer)

You Build Things That Think and Move

The robot tips forward.
Then it freezes.
Everyone watching holds their breath.
You kneel down, not worried—just curious.
"Okay," you say. "It didn't fall because it's bad. It fell because something's off."

- Maybe the wheels are too small.
- Maybe the weight isn't balanced.
- Maybe the instructions came in the wrong order.
- You change one tiny thing.
- The robot rolls again.
- This time, it keeps going.
- That moment when a machine finally moves the way you imagined is what robotics engineers live for.

This Might Be You If...What Data Detectives Really Do

- You like building things with your hands
- You enjoy fixing what's broken
- You don't mind trying again when something fails
- You're curious about how parts work together
- You get excited by buttons, wheels, wires, or moving parts

You might be the kid who:

- loves LEGO sets
- builds things from cardboard and tape
- asks, "What if it could move on its own?"

That's not just playing.
That's robotics thinking.

What Robotics Engineers Really Do

Robotics engineers design and build robots.

Robots are machines that can:

- move
- sense what's around them
- follow instructions
- sometimes make simple decisions

Some robots:

- clean floors
- explore space
- help doctors during surgery
- rescue people after disasters

Robotics engineers don't build a robot once and walk away. They:

- test it
- watch what goes wrong
- fix it
- test again

When a robot falls over, it's not a failure.
It's feedback.

How It Feels to Think Like This

Thinking like a robotics engineer feels hands-on and curious. It feels like:

- turning ideas into real objects
- watching something you built actually move
- learning from mistakes instead of getting upset
- feeling proud when a small change makes a big difference

Robotics engineers expect things to go wrong at first.
That's normal.
Every mistake teaches the robot—and the builder—something new.

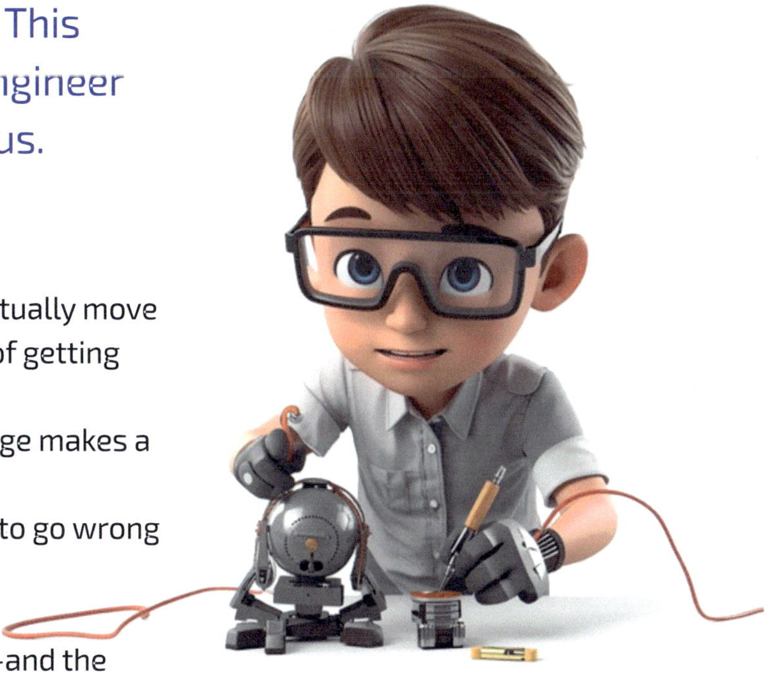

Tools You Might Use (and Why They Matter)

Robotics engineers use tools like:

- motors to help robots move
- sensors to help robots "see," "hear," or feel
- code to tell robots what to do
- computers and building kits

These tools matter because robots can go where humans can't. They can:

- explore deep oceans
- travel to space
- help people with disabilities
- do jobs that are too boring, dangerous, or hard for people

Robots don't replace humans.
They help humans do more.

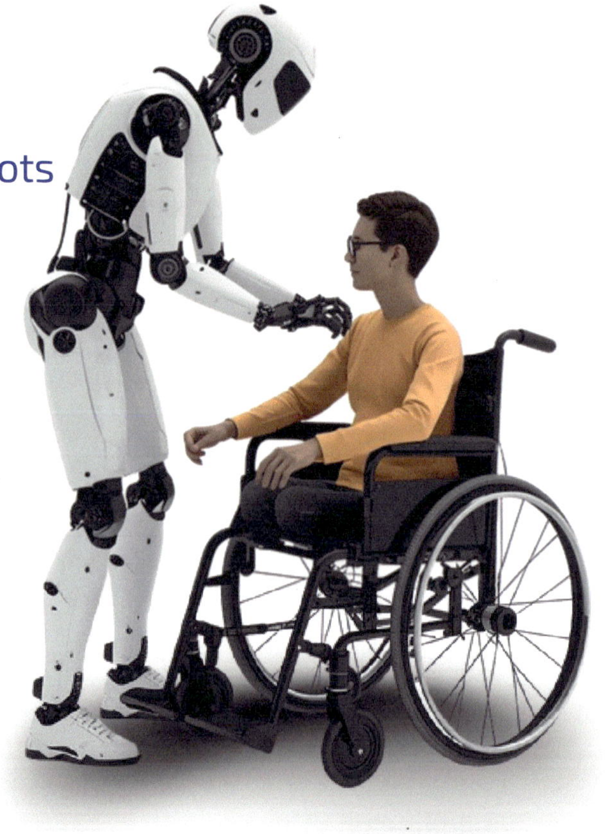

Try It Yourself

Design a Robot Pet

Design a robot pet on paper.

Think about:

• How does it move? Wheels, legs, or something else?

• What does the robot sense? Light, sound, touch?

• What should it do when it senses something?

Draw it and label the parts.

You're not just drawing.

You're planning how it works—exactly how robotics engineers begin.

One Last Thing to Know

You don't need fancy tools or perfect ideas.
What matters is that you:

- like building and testing
- don't give up when things fail
- enjoy learning by doing

Robotics engineers aren't just smart.

They're patient.
If you love creating things that move, learn, and help people...
This might be your kind of world.
You Might Also Like

Software Creator (Chapter 1) and Inventor & Prototyper (Chapter 16)

6. World Builder (Civil Engineer)

You Build the World People Live In

Every day, thousands of people cross a bridge.
Cars roll. Bikes pass. People walk and talk.
No one stops to think, Will this hold?
Because you already did.
If you've ever looked at a road, a building, or a bridge and wondered how it was made—or how it stays standing—you might already be thinking like a civil engineer.

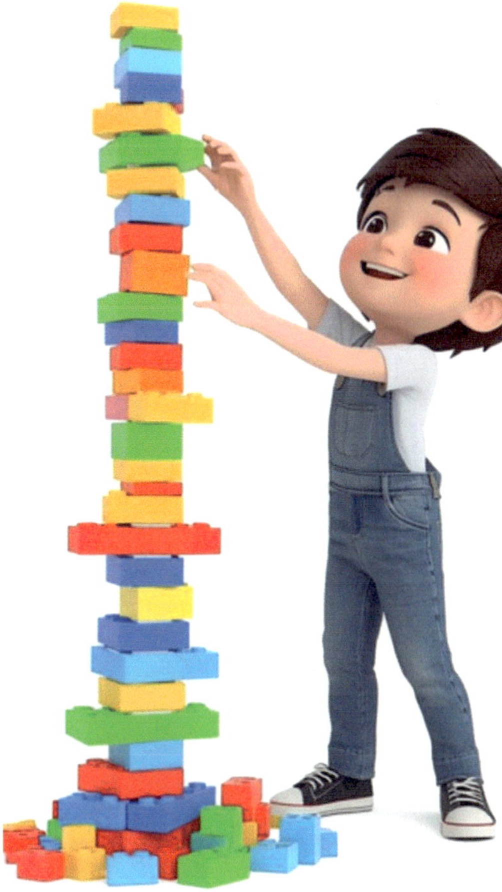

This Might Be You If...

You like big projects
You think about safety and strength
You enjoy planning before building
You like working with others
You care about things lasting a long time

You might be the kid who builds the tallest tower.
Or the one who checks if something is steady.
Or the one who asks, "Will this still work years from now?"
That kind of thinking builds cities.

What Civil Engineers Really Do

Civil engineers design and help build things people use every day.
They work on:

- Roads
- Bridges
- Schools
- Parks
- Buildings
- Water systems

Their job is to make sure these things are:

- Safe
- Strong
- Useful
- Built to last

Before anything is built, civil engineers plan carefully.
They study the land.
They think about the weather.
They calculate weight and pressure.
They work with teams to make smart choices.

A small mistake can cause big problems so civil engineers think ahead.

How It Feels to Think Like This

Thinking like a civil engineer feels responsible and steady.
It feels like:

- Knowing that people depend on your work
- Feeling proud of something that stands strong
- Caring about details that keep others safe
- Working patiently toward a long-term goal

Civil engineers don't rush.
They know strong things take time.

Tools You Might Use (and Why They Matter)
Civil engineers use tools like:

- Blueprints and maps
- Measuring tools
- Computer models
- Math and physics

These tools matter because cities don't build themselves.
Safe roads help people get where they need to go.
Strong bridges connect communities.
Clean water systems keep people healthy.
Civil engineers help everyday life run smoothly—even when no one notices.

Try It Yourself

Build a bridge using paper, tape,
or straws.
Test it by placing small objects on top.

Build a bridge using paper, tape, or straws.
Test it by placing small objects on top.

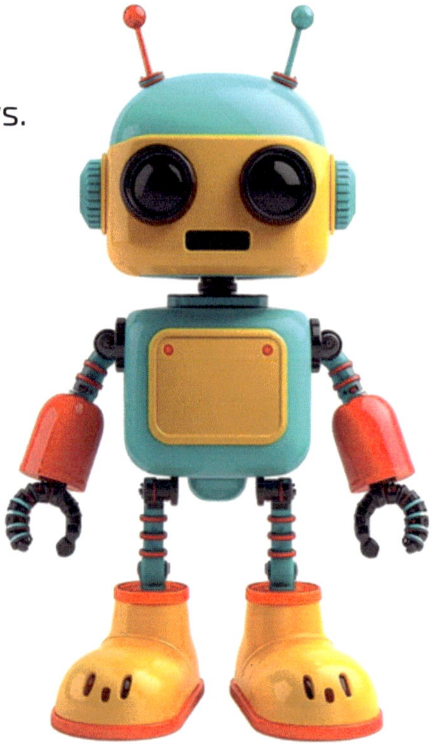

Ask yourself:

- What makes it stronger?
- Where does it bend or break?
- How could you improve it?

That's how engineers learn—by testing and improving.

One Last Thing to Know

You don't need to love being in the spotlight.
Civil engineers are often behind the scenes.

What matters is that you:

- Care about safety
- Think long-term
- Enjoy building things that help others

If you like creating strong, useful things people rely on every day…
This world might be built for you.

7. Planet Protector (Environmental Engineer)

You Fix Problems Humans Created

The creek behind the park looks different today.
The water is cloudy.
There's a strange smell.
No frogs. No fish.
People walk past and shake their heads.
"That's too bad," someone says.
You stop.
You don't just feel sad you feel curious.
What changed?
Where did the pollution come from?
And how do we fix it?
That moment when a problem doesn't just upset you, but makes you want to solve it is where environmental engineers begin.

This Might Be You If...

- You care about the planet
- You don't like waste or pollution
- You want to fix problems, not just talk about them
- You think about how today's choices affect tomorrow
- You believe humans can do better

You might be the kid who:

- notices trash on the ground
- asks where water really comes from
- thinks, "There has to be a smarter way"

That's not complaining.
That's problem-solving.

What Environmental Engineers Really Do

What Environmental Engineers Really Do
Environmental engineers design systems that protect both people and the planet.

They work on:

- cleaning drinking water
- reducing air pollution
- managing waste safely
- protecting soil, rivers, and ecosystems

They don't just study problems.

They fix them.
Environmental engineers use science and engineering to:

- clean up the damage humans have caused
- prevent future harm
- design safer, cleaner systems for cities and communities

They work with scientists, builders, and governments to turn good ideas into real solutions.

How It Feels to Think Like This

Thinking like an environmental engineer feels purposeful.
It feels like:

- wanting your work to truly matter
- caring about people, animals, and nature at the same time
- balancing progress with responsibility
- feeling hopeful instead of helpless

Environmental engineers don't panic when problems are big.
They break them into parts—and start fixing them.

Tools You Might Use (and Why They Matter)
Environmental engineers use tools like:

- filters that clean water and air
- testing equipment to measure pollution
- computer models to predict future problems
- renewable energy systems

These tools matter because clean air and water aren't optional. They protect:

- human health
- wildlife
- future generations

Environmental engineers help make sure the planet can support life—not just today, but long into the future.

Try It Yourself

Design a Cleaner System
Imagine your town or school.

Ask yourself:

• Where does trash go?
• Where does water come from
and where does it go after?
• Where could pollution sneak in?
Draw a simple path showing:
• the problem spot
• one change that could make things
cleaner

You don't need the perfect answer. Environmental engineering starts by noticing where systems can improve.

One Last Thing to Know

You don't have to be perfect to protect the planet. Environmental engineers aren't here to blame people. They're here to build better systems.

What matters is that you:

- *care about the world*
- *want practical solutions*
- *believe change is possible*

If you want to help people live better lives without harming the planet... This might be your kind of work.

You Might Also Like
Earth Scientist (Chapter 12) and Renewable Energy Designer (Chapter 14)

8. Sky Explorer (Aerospace Engineer)

You Help Humans Leave Earth

A rocket stands tall on the launch pad.
The ground shakes.
The countdown begins.
5...
4...
3...

Your heart beats faster—not because you're scared, but because you're curious. Will it work?

Did we calculate everything right?
Is this the moment humans go somewhere new?
If you look up at the sky and feel questions forming in your mind, you might already be thinking like an aerospace engineer.

This Might Be You If... You dream big

You're curious about flight and space
You like precision and accuracy
You enjoy solving difficult problems
You wonder what's beyond what we can see
You might be the kid who stares at airplanes.
Or the one who loves rockets, planets, and stars.
Or the one who asks, "How does that stay up there?"
That curiosity is powerful.

What Aerospace Engineers Really Do

Aerospace engineers design things that fly.

They work on:

- Airplanes
- Helicopters
- Rockets
- Satellites
- Spacecraft

Their job is to make sure these machines are:

- Safe
- Strong
- Efficient
- Able to handle extreme conditions

In space, there's no air to breathe and no room for mistakes.

Aerospace engineers think carefully about:

- Speed
- Weight
- Pressure
- Heat
- Balance

Even a tiny mistake can cause big problems—so they plan, test, and test again.

How It Feels to Think Like This

Thinking like an aerospace engineer feels exciting and serious at the same time.

It feels like:

- Wanting to explore the unknown
- Feeling amazed by science and physics
- Taking responsibility for human safety
- Working patiently toward something extraordinary

Aerospace engineers don't rush.
They know the sky—and space—demands respect.

Tools You Might Use (and Why They Matter)

Aerospace engineers use tools like:

- Computer simulations
- Wind tunnels
- Blueprints and models
- Math and physics

These tools matter because flying is complicated.
Airplanes connect people across the world.
Satellites help us communicate and predict the weather.
Spacecraft help us learn about our planet and beyond.
Aerospace engineers help humans explore safelyand smarter.

Try It Yourself

Test Tiny Changes (BUILD + TEST)
Make two paper airplanes.

Rules:

• Change only one thing between them (wing size, nose fold, or tail shape)
Test both airplanes.
Ask yourself:
• Which one flies farther?
• Which one flies straighter?
• What did the small change affect?
Aerospace engineers learn by testing tiny differences—one at a time.

One Last Thing to Know

You don't need to know everything about space right now.

What matters is that you:

• Ask big questions

• Care about accuracy

• Enjoy learning how things work

Aerospace engineers aren't just dreamers.
They're careful thinkers who turn dreams into reality.
If you love exploring, imagining, and pushing beyond limits...
The sky might not be the limit for you.

You Might Also Like

Robotics Engineer (Chapter 5) and Earth Scientist (Chapter 12)

SCIENCE CAREERS (4)

9. Truth Seeker (Research Scientist)

You Ask Questions That Change What We Know

Two plants sit on the same windowsill.
They get the same water.
The same soil.
The same care.
But one grows taller.
Most people shrug.
You don't.
You lean closer and think,
What's different?
The light? The air? The timing?
That moment when something doesn't match what you expected is where research scientists begin.

This Might Be You If...

- You ask a lot of "why" and "how" questions
- You like testing ideas instead of guessing
- You're patient and don't rush for answers
- You enjoy discovering something new
- You're okay being wrong if it helps you learn

You might be the kid who:

- runs little experiments at home
- notices when something doesn't make sense
- says, "Let's check," instead of arguing

That's not a doubt.
That's scientific thinking.

What Research Scientists Really Do

Research scientists study how the world works by asking careful questions.
They don't just collect facts.
They look for evidence.

Research scientists:

- ask clear questions
- make careful guesses (called hypotheses)
- design experiments to test those ideas
- observe what really happens
- learn from the results

Sometimes they're right.
Sometimes they're wrong.
Both matter.

In science, being wrong isn't failure it's information.

Research scientists work in many areas:

- *nature and the environment*
- *medicine and health*
- *energy and materials*
- *technology and space*

Anywhere there's something unknown, science belongs.

How It Feels to Think Like This

Thinking like a research scientist feels calm and focused.

It feels like:

- wanting the truth, even if it surprises you
- enjoying careful observation
- feeling excited when results finally make sense
- knowing answers take time—and that's okay

Research scientists don't rush.

They care more about being accurate than being fast.

Tools You Might Use (and Why They Matter)

Research scientists use tools like:

- *experiments and tests*
- *microscopes and measuring tools*
- *notebooks to record results*
- *labs, outdoor spaces, or even kitchens*

These tools matter because science helps people understand reality.

Science helps:

- *doctors treat illness*
- *farmers grow food*
- *engineers build safely*
- *communities solve problems*

Without science, we'd still be guessing.

Try It Yourself

Test an Idea (OBSERVE + TEST)

Choose one question to test:

- Which melts ice faster: metal or plastic?
- Which dries faster: paper towel or cloth?
- Which dissolves faster: sugar in warm water or cold water?

Before you start, write your prediction. Then test it.

Afterward, write:

- what actually happened
- what surprised you

That's real science—not memorizing answers but learning from results.

One Last Thing to Know

You don't need to be loud or flashy to be a scientist.

Many research scientists are:

- quiet
- curious
- thoughtful
- careful

What matters most is that you:

- ask honest questions
- look for real answers
- care about learning what's true

If you enjoy noticing when things don't match expectations—and figuring out why...
Science might already be part of who you are.

You Might Also Like

Data Detective (Chapter 4) and Chemist (Chapter 10)

10. Matter Mixer (Chemist)

You Rearrange Matter to Create Something New
Imagine This...

Two things look ordinary on their own.
Then they mix.
Suddenly, there's a color change.
A fizzing sound.
Heat.
A new smell.
Something new has been created.
If you've ever mixed things just to see what would happen—and watched closely instead of walking away—you might already be thinking like a chemist.

This Might Be You If...

You're curious about how things change
You like following rules and steps
You enjoy reactions and results
You notice details others miss
You want to know why something happens
You might be the kid who loves slime experiments.
Or the one who reads ingredient labels.
Or the one who asks, "What happens if we add this?"

That curiosity is powerful.

What Chemists Really Do

Chemists study matter—what things are made of and how they change. Everything around you is made of matter:

- Air
- Water
- Food
- Clothes
- Medicine

Chemists look at matter on a tiny scale and ask:

- What is this made of?
- How does it react with other things?
- How can we change it safely?

Chemists help create:

- Medicine that saves lives
- Cleaners who keep places safe
- Materials that are stronger or lighter
- Food products that last longer

They don't mix things randomly.
They follow rules carefully, because small changes can cause big reactions.

How It Feels to Think Like This

Thinking like a chemist feels focused and careful.

It feels like:

- Watching closely and noticing changes
- Feeling excited when a reaction works
- Respecting rules that keep everyone safe
- Learning from both successes and mistakes

Chemists know that safety matters.
They plan first.
Then they test.

Tools You Might Use (and Why They Matter)

Chemists use tools like:

- Lab equipment, like beakers and test tubes
- Measuring tools
- Formulas and charts
- Protective gear

These tools matter because chemistry affects daily life.

Chemistry helps:

- Doctors treat patients
- Farmers protect crops
- Engineers build better materials
- Communities stay healthy

Chemistry turns ideas into useful solutions.

Try It Yourself

Chemical Change vs Physical Change

You'll need:

- A piece of paper
- A pencil
- A cup of water
- A spoon
- A pinch of salt or sugar

Step 1: Physical Change

Tear the paper in half.

Ask yourself:

- Is it still paper?
- Could you put it back together?

That's a physical change the material didn't turn into something new.

Step 2: Chemical Change (Safe Version)

Put a small amount of salt or sugar into the water.

Stir it.

Ask yourself:

- Can you still see the sugar or salt?
- Did it disappear, or just change how it looks?

Chemists ask questions like this all the time.

Step 3: Think Like a Chemist

Now answer these questions:

- Which change was easier to reverse?
- Which one changed how the material behaved?
- Why do you think chemists care about this difference?

You just practiced chemical thinking — noticing how matter changes, and why it matters.

One Last Thing to Know

You don't need a lab to start thinking like a chemist.

What matters is that you:

- Like understanding change
- Follow the steps carefully
- Respect safety and rules

Chemists aren't just mixers.
They're careful thinkers who help create things people depend on every day.
If you enjoy watching how small changes lead to big results...
This might be your kind of science.

You Might Also Like
Research Scientist (Chapter 9) and Medical Technology Innovator (Chapter 15)

11. Life Explorer (Biologist)

You Study Life Itself

Your body is home to more bacteria than human cells.
Most of them are not dangerous.
Many of them actually help keep you alive.
Biologists know this—which is why they don't just look at life.
They study how it works together.

This Might Be You If...

- You like watching animals or plants closely
- You're curious about how bodies grow, heal, or survive
- You enjoy nature, even the small or "gross" parts
- You're patient and notice tiny details
- You care about living things

You might be the kid who:

- watches ants instead of stepping over them
- wonders why leaves change color
- asks how muscles, bones, or organs work

That's biology thinking.

What Biologists Really Do
Biologists study living things.

That includes:

- plants
- animals
- humans
- cells
- ecosystems

They ask questions like:

- How do living things grow?
- How do they survive in different environments?
- How do living things depend on each other?
- Some biologists work in labs.
- Some work outdoors.
- Some study humans.
- Some study animals, plants, or tiny cells.
- All of them are trying to understand life better—not just for curiosity, but to protect it.

How It Feels to Think Like This

Thinking like a biologist feels calm and observant.

It feels like:

- slowing down to really notice
- being curious instead of rushing
- feeling amazed by small details
- wanting to protect living systems

Biologists don't guess.

They watch.

They compare.

They learn over time.

Tools You Might Use (and Why They Matter)

Biologists use tools like:

- microscopes to see tiny life
- notebooks for observations
- cameras and measuring tools
- lab or field equipment

These tools matter because understanding life helps people.

Biology helps:

- doctors treat illness
- farmers grow food
- conservationists protect wildlife
- scientists understand ecosystems

When we understand life better, we take better care of it.

Try It Yourself

Observe Like a Biologist (OBSERVE)

Pick one living thing and observe it quietly for 10 minutes:

- an insect
- a plant
- a pet
- birds outside your window

Write or draw:

- how it moves (or doesn't)
- what it reacts to
- one pattern you notice

Now ask yourself:

- Why might it behave that way?

That's exactly how biologists start.

One Last Thing to Know

You don't need to travel to a rainforest to be a biologist.

Many biologists are:

- thoughtful
- curious
- patient
- caring

What matters is that you:

- respect life
- enjoy learning how living things work
- want to understand nature instead of ignoring it

If you love observing life and asking quiet questions about how living systems survive... Biology might already be part of who you are.

You Might Also Like
Earth Scientist (Chapter 12) and Medical Technology Innovator (Chapter 15)

12. Planet Reader (Earth & Climate Scientist)

You Read the Planet's Clues

The rain was heavy last night.
In the morning, one street is flooded.
Another street nearby is completely dry.
Same storm.
Different outcome.
An earth scientist looks at the ground, the slope, the soil, and the drains—and starts to understand why.

This Might Be You If...

- You like learning about Earth, weather, and land
- You notice patterns in nature
- You ask questions about storms, earthquakes, or volcanoes
- You enjoy big systems that connect many parts
- You want to understand how the planet changes

You might be the kid who:

- studies maps or globes
- watches the sky before a storm
- asks why some places flood or shake more than others

That curiosity is Earth science thinking.

What Earth Scientists Really Do Earth scientists study the planet we live on.

They learn about:

- rocks and soil
- mountains and valleys
- rivers and oceans
- weather and climate
- natural events like earthquakes, floods, and volcanoes

Earth scientists don't just study the past.

They use clues to understand what might happen next.

They help communities:

- prepare for storms
- understand climate change
- build safer cities
- protect natural resources

Their job is to listen to what the planet is telling us—and help people respond wisely

How It Feels to Think Like This

Thinking like an earth scientist feels observant and patient. It feels like:

• noticing slow changes over time
• seeing connections between land, water, and air
• respecting nature's power
• wanting to help people stay safe
Earth scientists know the planet is always changing.
Their job is to pay attention.

Tools You Might Use (and Why They Matter)

Earth scientists use tools like:

• maps and globes
• satellites that watch Earth from space
• weather tools and sensors
• rock and soil samples
These tools matter because understanding Earth saves lives.
Earth science helps:
• predict storms and floods
• plan safer buildings
• protect water sources
• understand climate change

When we read Earth's clues correctly, we make better choices.

Try It Yourself

Read the Land (DESIGN + OBSERVE)

Draw a simple map of your area—or an imaginary town. Include:

- roads
- buildings
- trees or grass
- places where water might flow

Now answer:

- Where would rainwater collect?
- Which areas might flood first?
- What one change could help water drain better?

You just practiced earth science thinking—reading the land and planning ahead.

One Last Thing to Know

You don't need to travel the world to be an earth scientist.

Many earth scientists work by:

- studying maps and data
- watching patterns over time
- asking careful questions

What matters is that you:

- care about the planet
- like big-picture thinking
- want to understand how Earth supports life

If you enjoy reading nature's clues and helping people live safely on this planet…
This might be your kind of science.

You Might Also Like
Environmental Engineer (Chapter 7) and Renewable Energy Designer (Chapter 14)

FUTURE & HYBRID STEM (4)

13. Machine Teacher (AI Engineer)

You Teach Machines How to Learn

A computer helps choose who gets extra help in school.
But what if the computer learned from unfair examples?
What if it starts helping some kids more than others—without anyone meaning to?
Do you turn it off?
Will you fix it?
Do you change the rules?
That's the kind of decision an AI builder has to make.

This Might Be You If...

- You like figuring out how rules work
- You enjoy patterns and logic
- You ask "What if?" a lot
- You care about fairness and responsibility
- You wonder how machines make decisions

You might be the kid who:

- asks why an app suggests certain videos
- likes teaching others how to do things step by step
- wonders whether computers can be trusted.

That's AI thinking.

What AI Builders Really Do

AI builders create systems that can learn from information.

Instead of giving a computer every instruction, they:

- set rules
- show examples
- let the system find patterns

AI systems help:

- translate languages
- recommend videos or music
- recognize speech or images
- help doctors make decisions

But AI builders don't just ask, "Can we build this?"

They ask, "Is this fair? Is this safe? Is this helpful?"

Because machines learn from humans and humans make mistakes

How It Feels to Think Like This

Tools You Might Use (and Why They Matter)

AI builders use tools like:
- data sets (examples the computer learns from)
- algorithms (the rules for learning)
- coding languages
- testing tools to check fairness and accuracy

These tools matter because AI is becoming part of everyday life.

AI helps:
- doctors diagnose illness
- students learn in new ways
- cars drive more safely
- people communicate across languages

Good AI helps people.
Careless AI can cause harm.

Try It Yourself

Build Rules, Then Fix Them (SOLVE)
Create a simple decision system for snacks.

Rules:

- If someone dislikes chocolate → recommend fruit
- If someone is allergic to nuts → avoid nuts

Now answer:

- What if two people are very different?
- Is it fair to recommend the same snack to everyone?

Add one fairness rule to improve your system.
That's how AI builders think by adjusting rules to improve outcomes.

One Last Thing to Know

You don't need to be a computer genius to think like an AI builder.

What matters is that you:

- like understanding decisions
- care about fairness
- think about consequences

AI builders aren't just programmers.
They're careful thinkers shaping how technology treats people.
If you enjoy teaching systems how to learn—and making sure they learn the right way...
This might be your kind of future.

You Might Also Like
Data Detective (Chapter 4) and Cyber Defender (Chapter 3)

14. Future Power Builder (Clean Energy Engineer)

You Power the Future

The sun gives Earth more energy in one hour than humans use in an entire year.
The problem isn't the energy.
It's figuring out how to catch it, store it, and use it when we need it.
That's the job of a renewable energy designer.

This Might Be You If...

You think about where electricity comes from
You care about the environment and people's health
You like building smart solutions
You imagine a better future
You ask, "Is there a cleaner way to do this?"
You might be the kid who notices solar panels on rooftops.
Or the one who wonders how windmills work.
Or the one who thinks, "There has to be a smarter way to power things."
That thinking matters more than ever.

You Power the Future

The sun gives Earth more energy in one hour than humans use in an entire year.
The problem isn't the energy.
It's figuring out how to catch it, store it, and use it when we need it.
That's the job of a renewable energy designer.

This Might Be You If...

You think about where electricity comes from
You care about the environment and people's health
You like building smart solutions
You imagine a better future
You ask, "Is there a cleaner way to do this?"
You might be the kid who notices solar panels on rooftops.
Or the one who wonders how windmills work.
Or the one who thinks, "There has to be a smarter way to power things."
That thinking matters more than ever.

What Renewable Energy Designers Really Do

Renewable energy designers create ways to power the world without running out of resources.
They work with energy from:

- The sun (solar power)
- The wind (wind power)
- Water (hydropower)
- Heat from inside the Earth (geothermal energy)

Their goal is to design systems that are:

- Clean
- Safe
- Reliable
- Affordable

They don't just build machines.
They plan how energy systems fit into cities, homes, and communities.
They think about:

- Where energy is needed
- How much power is required
- How to store energy
- How to protect nature at the same time

How It Feels to Think Like This

Thinking like a renewable energy designer feels hopeful.

It feels like:

- Caring about the future
- Wanting to solve real-world problems
- Balancing technology and nature
- Knowing your work can help many people

Renewable energy designers don't just think about today.
They think about tomorrow—and the years after that.

Tools You Might Use (and Why They Matter)

Renewable energy designers use tools like:

- Solar panels and wind turbines
- Energy storage systems like batteries
- Computer models and simulations
- Measuring and testing equipment

These tools matter because energy affects everything.

Energy powers:

- Homes
- Schools
- Hospitals
- Transportation

Clean energy helps protect the planet while keeping life running smoothly.

Try It Yourself

Design a Clean-Energy Building (DESIGN)
Design a school, home, or small town powered by clean energy.

On paper, label:

- where energy comes from
- how it's stored
- what happens at night or on calm days

Now answer:

- What problem was hardest to solve?
- What part are you most proud of?

That's renewable energy thinking planning systems, not just parts.

One Last Thing to Know

You don't need to invent everything from scratch.

Renewable energy designers:

• Improve existing ideas
• Test new solutions
• Work with teams

What matters is that you:

• Care about people and the planet
• Like solving practical problems
• Think long-term

If you want to help power the world in a cleaner, smarter way…
This might be your kind of future

You Might Also Like

Environmental Engineer (Chapter 7) and Civil Engineer (Chapter 6)

15. Health Helper (Biomedical Engineer)

You Combine Science and Compassion

Someone isn't feeling well.
They're scared.
They're waiting for help.
Now imagine being part of the reason that help exists.
A new tool makes the test faster.
A new idea helps a doctor see a problem sooner.
A new invention makes treatment safer, easier, or less painful.
If you care about people and like solving problems, you might already be thinking like a medical innovator.

This Might Be You If...

You care deeply about helping others
You're curious about how the human body works
You like improving things instead of accepting "that's how it is"
You think about comfort, safety, and fairness
You want your work to truly matter
You might be the kid who checks on others when they're hurt.
Or the one who asks how medicine works.
Or the one who thinks, "There has to be a better way to help."
That mix of kindness and curiosity is powerful.

What Medical Innovators Really Do

Medical innovators create new ways to help people stay healthy or get better. They don't all do the same job.

Some invent:

- Medical devices
- Tools doctors use during surgery
- Machines that help diagnose illness

Some improve:

- Treatments
- Testing methods
- Hospital systems

Some work on:

- New medicines
- Better vaccines
- Technology that helps doctors make decisions

Medical innovators work with doctors, scientists, engineers, and patients. They listen closely to problems and look for solutions that actually help real people. Their goal isn't just to make something new.

It's to make something useful.

How It Feels to Think Like This

Thinking like a medical innovator feels caring and focused.

It feels like:
• Wanting to ease pain or fear
• Feeling responsible for others' well-being
• Being careful because lives matter
• Feeling proud when your work helps someone heal

Medical innovators take their time.
They test carefully.
They double-check results.
They think about safety first.
Because helping people means doing things right.

Tools You Might Use (and Why They Matter)

Medical innovators use tools like:
• Medical machines and devices
• Computer models
• Lab tests and data
• Design tools for building new solutions

These tools matter because healthcare affects everyone.

Medical innovation helps:
• Doctors diagnose illness earlier
• Patients recover faster
• Hospitals work more efficiently
• Communities stay healthier
Behind every medical breakthrough is someone who cared enough to improve things.

Try It Yourself

Design for Comfort (DESIGN)

Think about a doctor or clinic visit.

Ask yourself:
- What felt confusing?
- What felt uncomfortable or scary?
- What took too long?

Now design one improvement that could:

- reduce fear
- save time
- improve comfort

Draw it or write about it.

Medical innovation often begins with empathy—seeing problems through someone else's eyes.

One Last Thing to Know

You don't need to be a doctor to help people heal.
Medical innovators come from many backgrounds.

What matters is that you:

- Care about others
- Like solving meaningful problems
- Respect science and safety

Medical innovators aren't just smart.
They're compassionate.

If you want to use science, technology, and creativity to help people live healthier lives...
This might be your kind of work.

16. Idea Maker (Product Designer / Inventor)

You Make What Doesn't Exist Yet

There's a problem everyone complains about.
"This is annoying."
"This takes too long."
"This never works right."
Then someone shrugs and says,
"That's just how it is."
You have a choice.
Do you accept it…
or do you try to build something better?
Inventors choose the second option.

This Might Be You If…

- You come up with ideas constantly
- You ask "What if?" more than "Why not?"
- You like building, tinkering, or experimenting
- You don't mind when ideas fail at first
- You enjoy solving problems in your own way

You might be the kid who:

- turns boxes into machines
- sketches ideas in notebooks
- suggests improvements no one asked for

That's not difficult.
That's inventing.

What Inventors & Prototypers Really Do

Inventors create new solutions to real problems.
Sometimes they:

- invent brand-new tools
- redesign objects that already exist
- improve systems that don't work well

They don't wait for perfect ideas.

They build prototypes—early versions that help them test ideas quickly.
Most inventions:

- it doesn't work the first time
- looks messy at first
- need many changes

That's normal.
Inventors expect mistakes.
They use them as clues.

How It Feels to Think Like This

Thinking like an inventor feels creative and brave.

It feels like:
• trusting your ideas
• feeling excited by possibilities
• learning from failure instead of fearing it
• improving something step by step
Inventors don't need permission. They start where they are, with what they have.

Tools You Might Use (and Why They Matter)

Inventors use many tools, depending on what they're creating:
• paper and pencils for sketches
• cardboard, tape, or recycled materials
• computers and software
• prototypes to test ideas
These tools matter because ideas need a way to become real.
Inventions help:
• make life easier
• improve safety
• solve everyday problems
• create new opportunities
Behind every useful object was someone who decided to try.

Try It Yourself

Prototype, Test, Improve (BUILD)

Pick a small problem in your daily life:

- tangled cords
- a messy desk
- a hard-to-open container

Using only household items (paper, tape, cardboard), build a rough solution.

Then:

- test it once
- improve it once

You don't need perfection.
You just made a prototype—and that's real inventing.

One Last Thing to Know

One Last Thing to Know
You don't have to be "the smartest person in the room."

Inventors are:

• curious • persistent
• creative • willing to try again

What matters most is that you:

• believe problems can be improved
• aren't afraid to experiment
• keep going when ideas don't work right away

Inventors shape the future by asking a simple question:
"What if there's a better way?"
If you love creating, experimenting, and bringing ideas to life...
This might be your world.

You Might Also Like
Game Designer (Chapter 2) and Robotics Engineer (Chapter 5)

You Don't Have to Choose—You Just Have to Notice

If you made it this far, there's something important you should know.
You don't need to have one answer.
You don't need to decide your future right now.
You don't need to pick one career and stick to it forever.
That's not how real lives work.

As you read this book, you may have noticed something. More than one chapter probably felt like you.

Maybe you saw yourself in:

- The kid who loves solving puzzles
- The one who builds things with their hands
- The one who cares deeply about people or the planet
- The one who asks questions that don't have easy answers

That doesn't mean you're confused.
It means you're curious.
And curiosity is not a problem to fix.
It's a strength to protect.

Some people try to put kids in boxes early.

"This one is a math kid."
"That one is creative."
"This one is practical."

But real thinkers don't fit neatly into boxes.
You can love science and art.
You can care about people and technology.
You can build things and imagine new worlds.

STEM isn't one narrow path.

It's a way of thinking.
It's about asking questions.
It's about noticing patterns.
It's about caring enough to solve problems.
It's about trying, failing, learning, and trying again.

Those skills don't disappear.
They grow with you.
As you get older, your interests may change.
Or they may deepen. Both are okay.

What matters is that you keep paying attention to
what excites your mind and pulls your curiosity forward.
If something made you think,
"I want to learn more about that."
That's a clue.

If something made time feel shorter,
That's another clue.
You don't have to act on every clue right now.
Just collect them.
Write them down.
Talk about them.
Explore them when you can.
Your future isn't a single choice waiting to trap you.
It's a series of doors.

And you don't have to open them all today.
You just have to notice which ones make you curious
enough to knock.

And remember this:

You don't need to become someone else to succeed.
You're already starting as you.
And that's exactly the right place to begin.

Explore Your Imagination!

Unlock a treasure trove of reading and coloring books!

1. Simply open your camera app on your phone.
2. Point the mobile device at the QR code
3. This will take you directly to my author central page.

or visit (http://www.letscolorimaginations.com) to delve into a vibrant world of creativity and fun.

Don't miss out—join our mailing list through the website to receive a free 22-page activity book, or download select books as MP3 files.

Happy reading and coloring !

Check out my other books at
www.letscolorimaginations.com

Magical World of Princess Coloring Book
(For Girls ages 4-8 years)

Magical World of Princess Coloring Book
(For Girls ages 8-12 years)

Princess, Unicorn, Mermaid, Fairy and Pegasus
Activity Book (For Kids ages 4-8) Volume 1, 2, 3

Princess, Unicorn, Mermaid, Fairy and Pegasus
Activity Book (For Kids ages 8-12) Volume 1, 2, 3

I Can Be Anything I Want To Be
Inspiring Careers for Girls
(Coloring Book)

Can Be Anything I Want To Be
Inspiring Careers for Boys
(Coloring Book)

Check out my other books at
www.letscolorimaginations.com

Unicorn Coloring Book (For Kids ages 4-8)

Unicorn Coloring Book (For Kids ages 8-12)

Mermaids Coloring Book (For Kids ages 4-8)

Mermaids Coloring Book (For Kids ages 8-12)

Magical World of Fairies Coloring Book
(For Kids ages 4-8)

Magical World of Fairies Coloring Book
(For Kids ages 8-12)

Enchanted Fairy World Coloring Book
(For Kids ages 8-12 & Adults)

Illustrated books

Grown up me ,will be
Inspiring careers for kids 4-8y/o

When I grow up I want to be.
Inspiring stem careers for kid 8-11y/o
Coloring books

Chibi girls coloring book

Zendoddle Designs
for kids 8-12
Careers Coloring Book

Thank you so much for purchasing my book!

Your support as a reader is invaluable, especially as a small indie publisher. Knowing that my book may be playing a role in your child's education is incredibly fulfilling.

If you could spare a minute to leave your honest feedback on Amazon, it would be immensely helpful. Your review not only aids in improving my future work but also guides other parents in their choice of educational resources.

Once again, thank you for being a part of this journey. Your thoughts and experiences with the book are truly appreciated!

To leave feedback

1. Open up your camera app.
2. Point your mobile device at the QR code below.
3. The review page will appear in your web browser.

or you can search on Amazon for the name of the book and Elizabeth Shelby

Made in the USA
Monee, IL
05 November 2026

67997222R10057